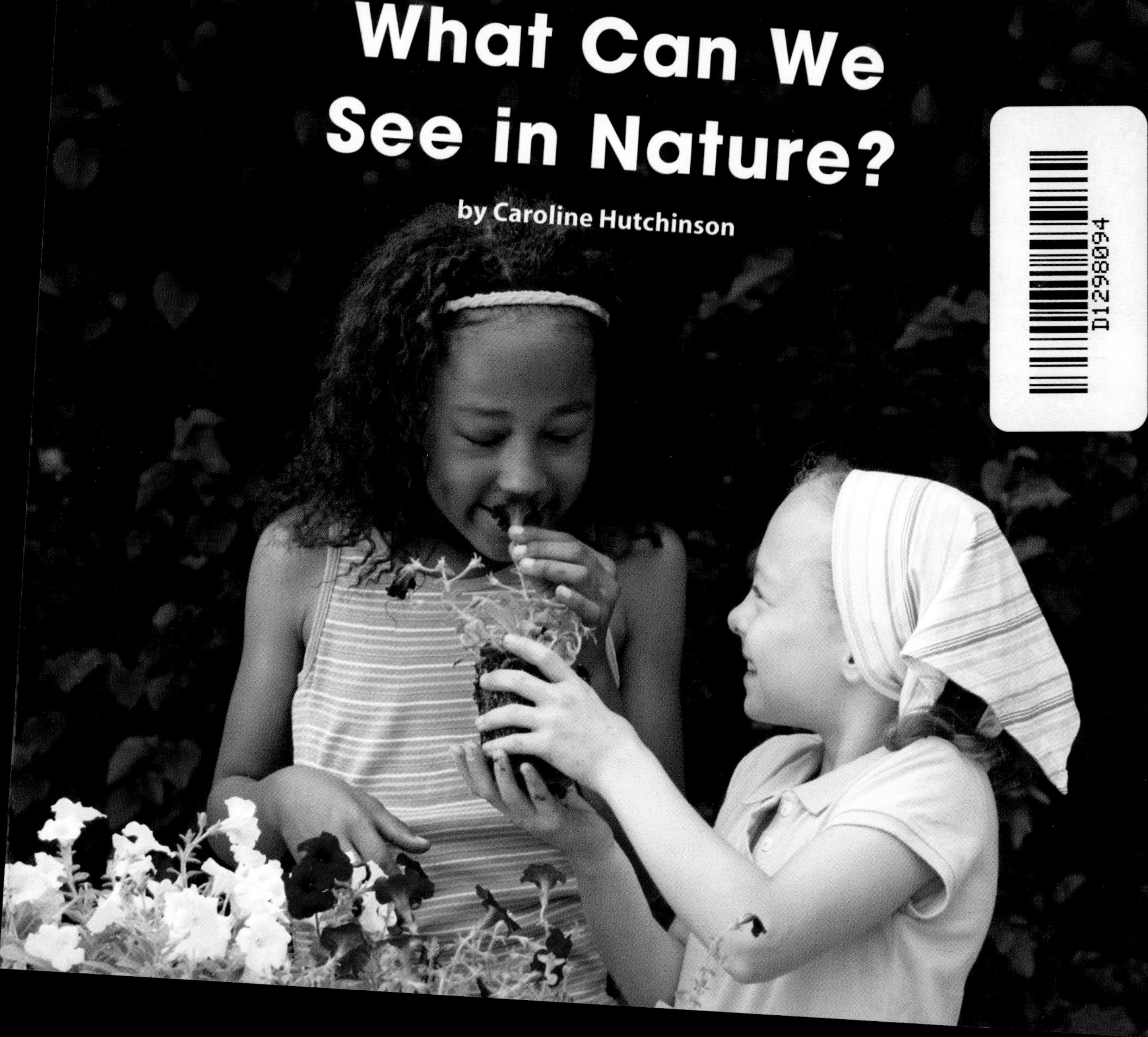

What Can We See in Nature?

by Caroline Hutchinson

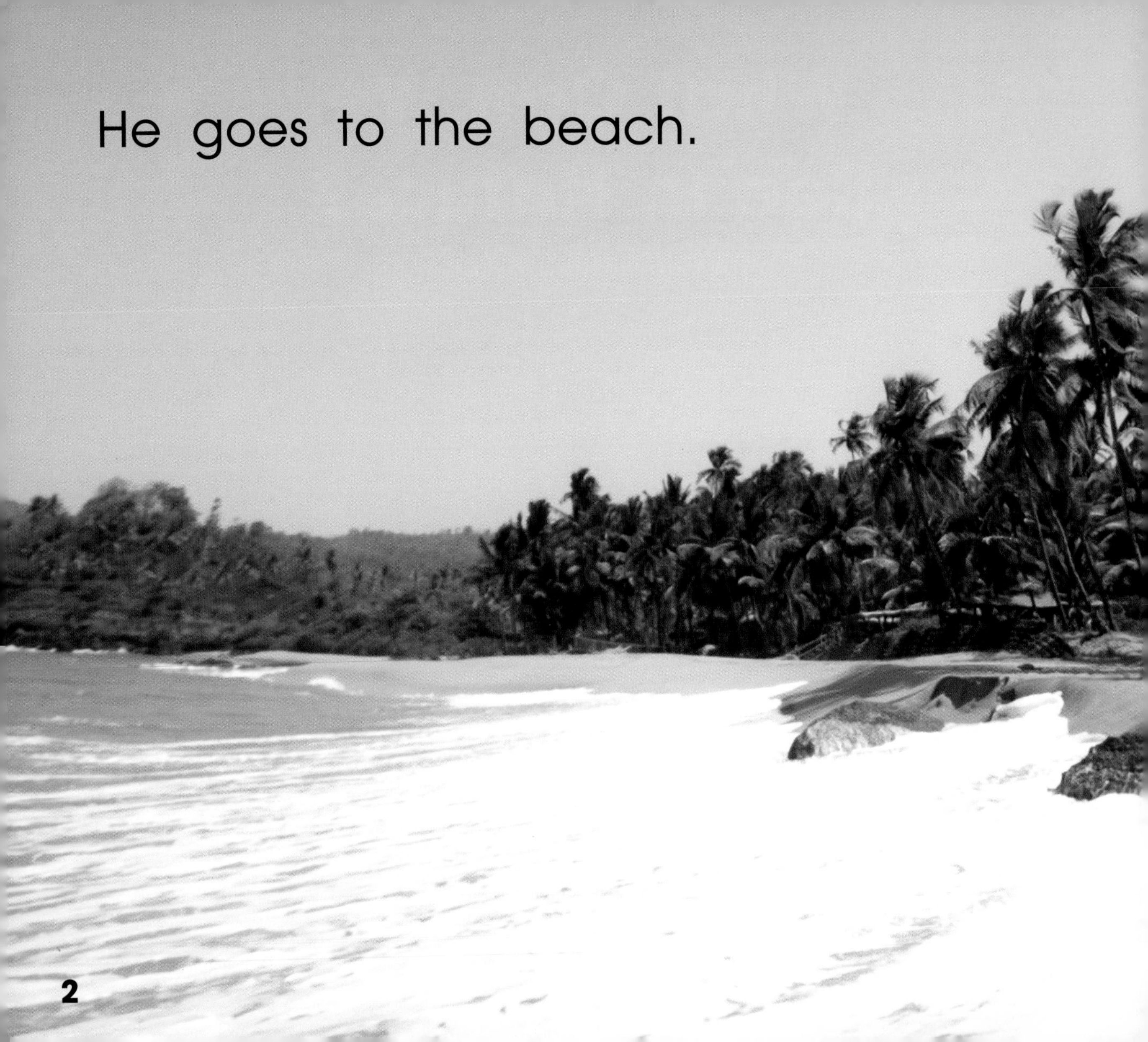

He goes to the beach.

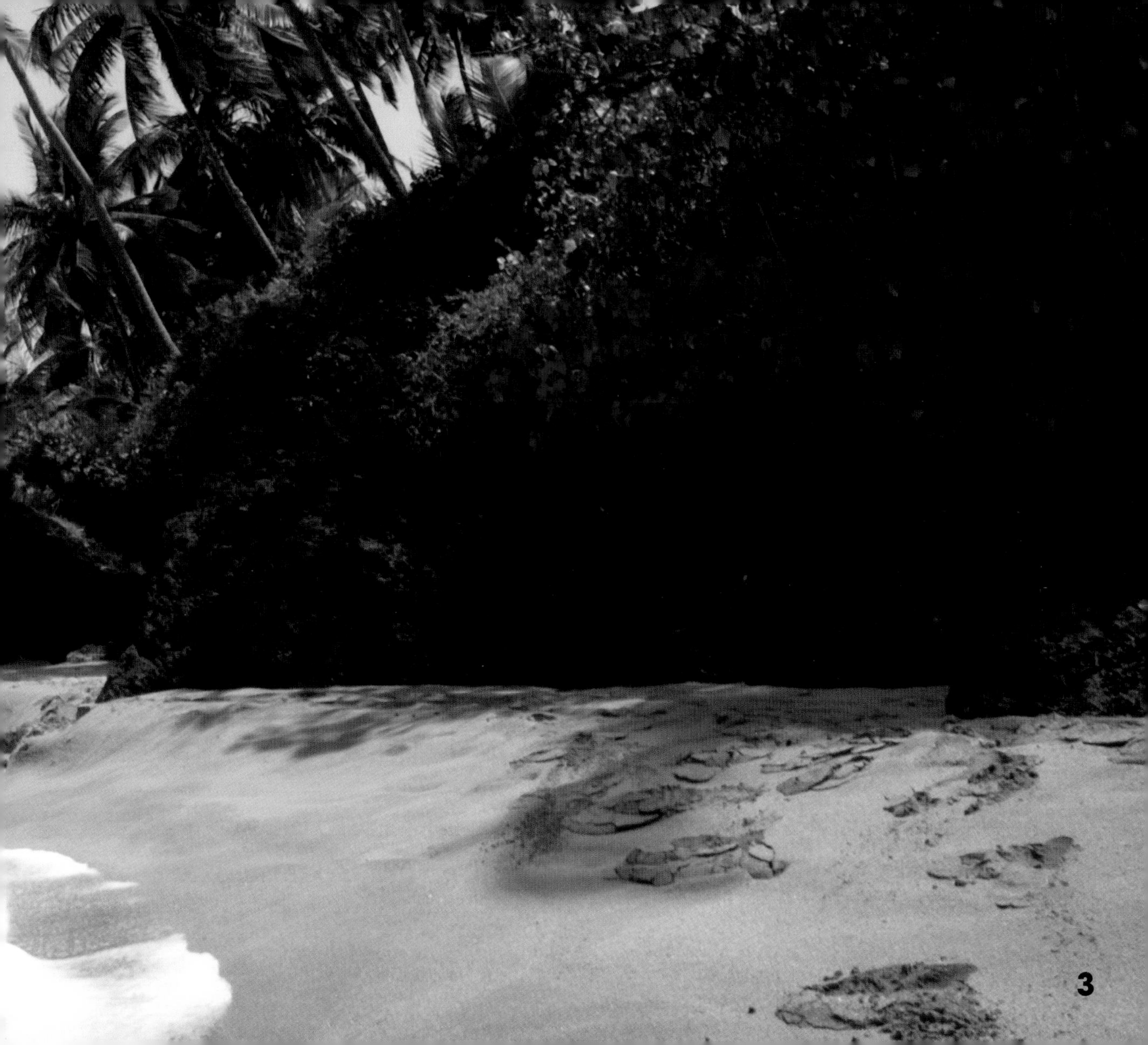

He likes to look at shells.

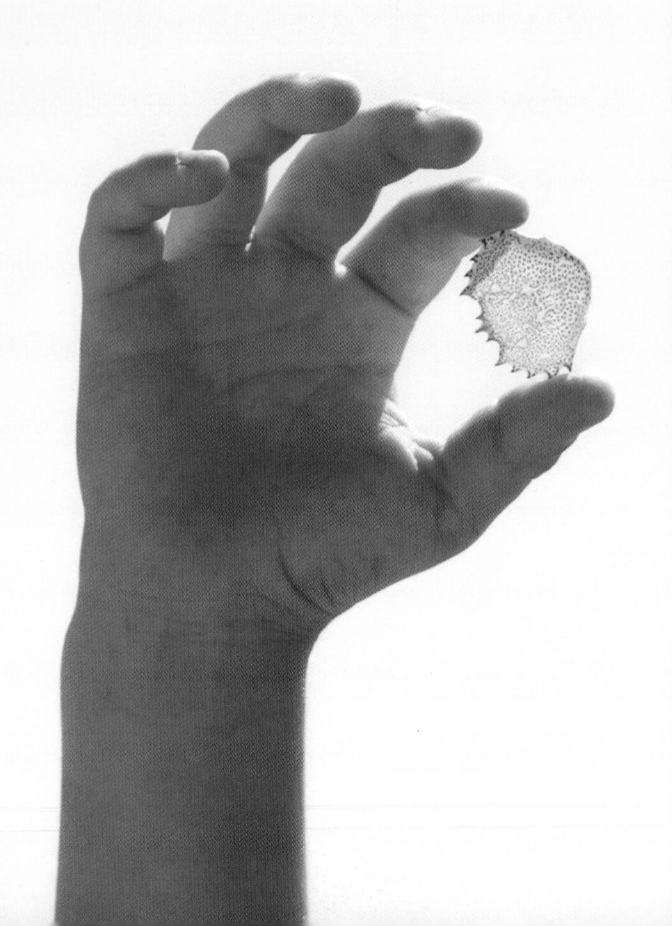

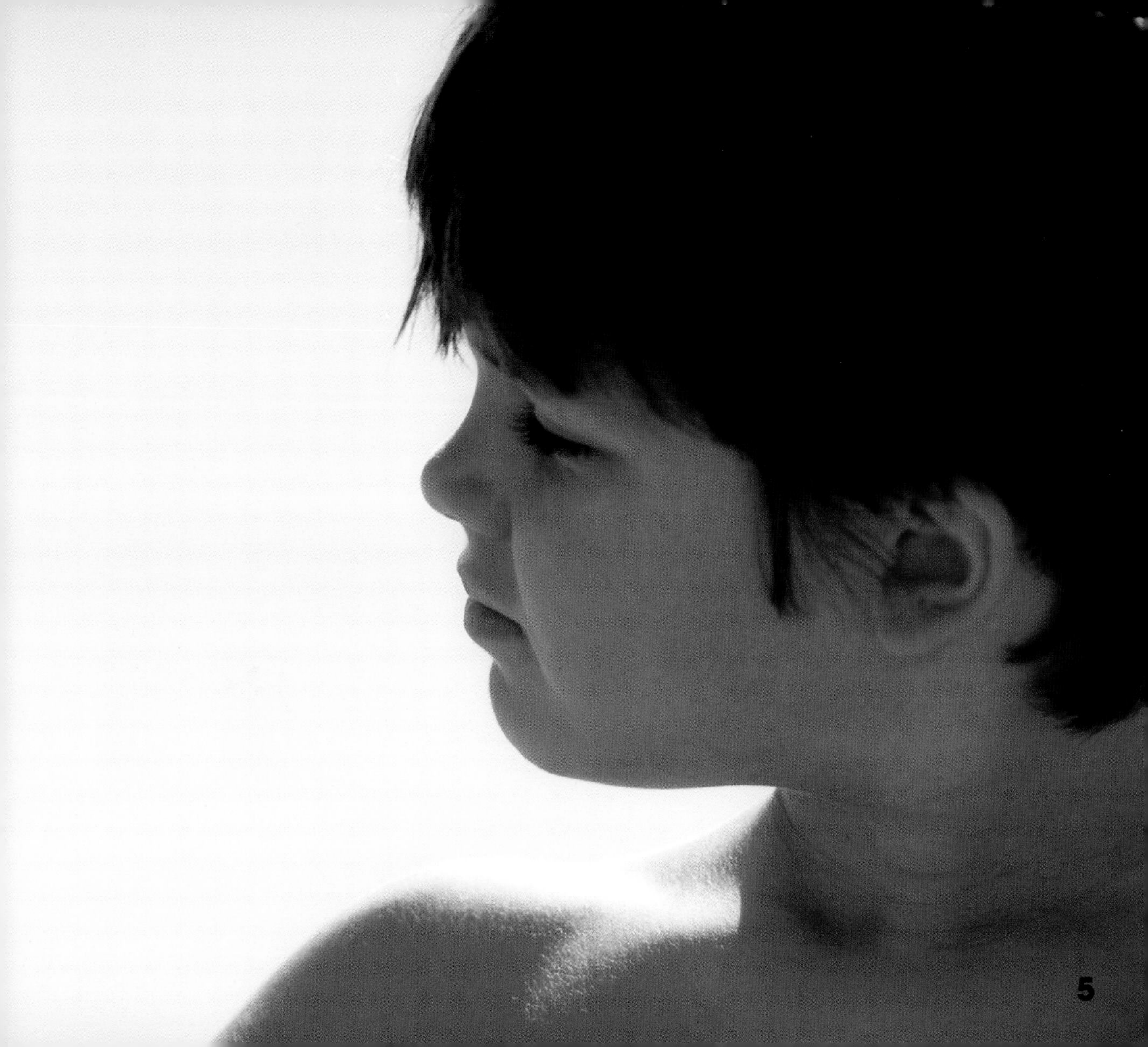

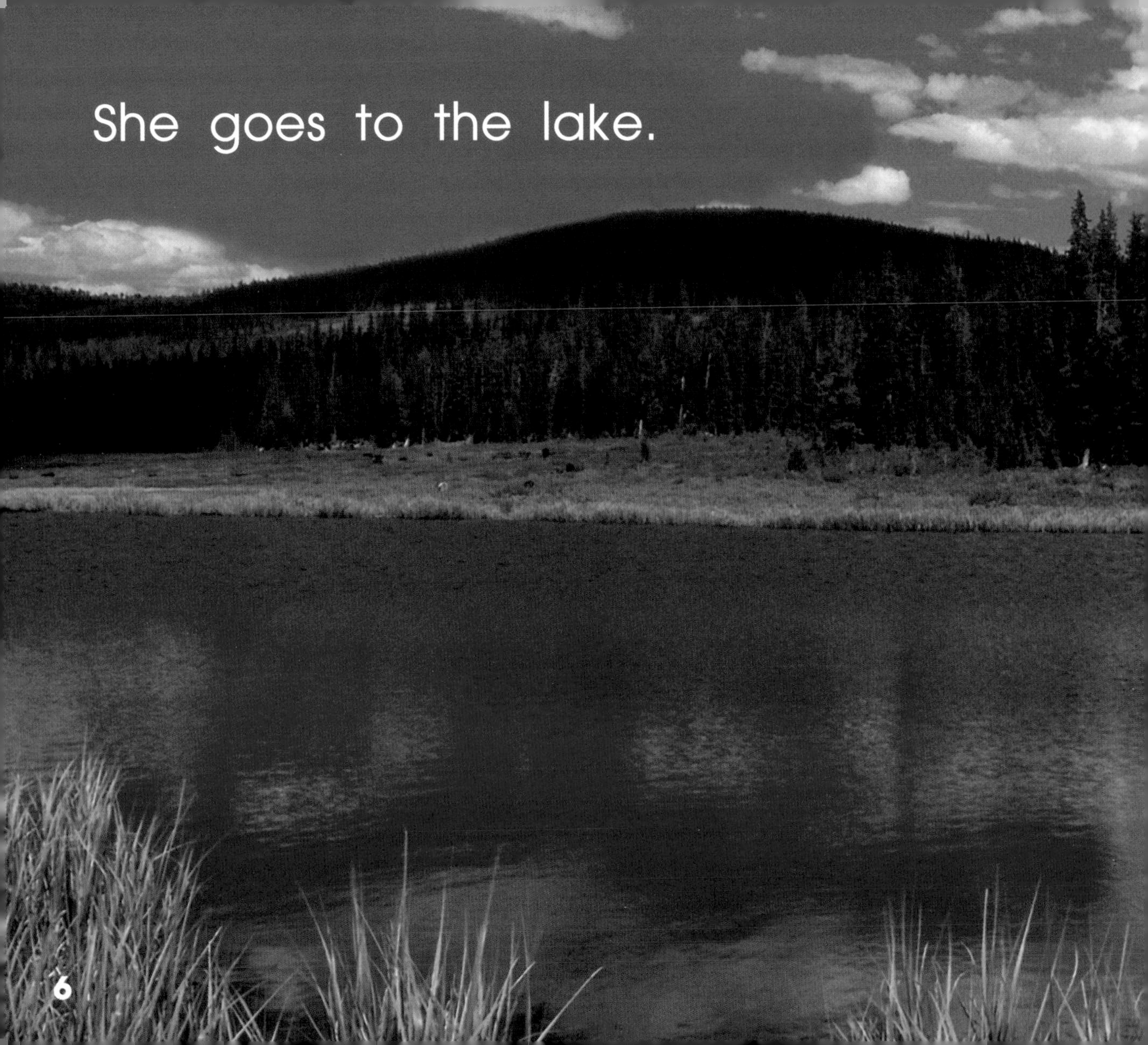

She goes to the lake.

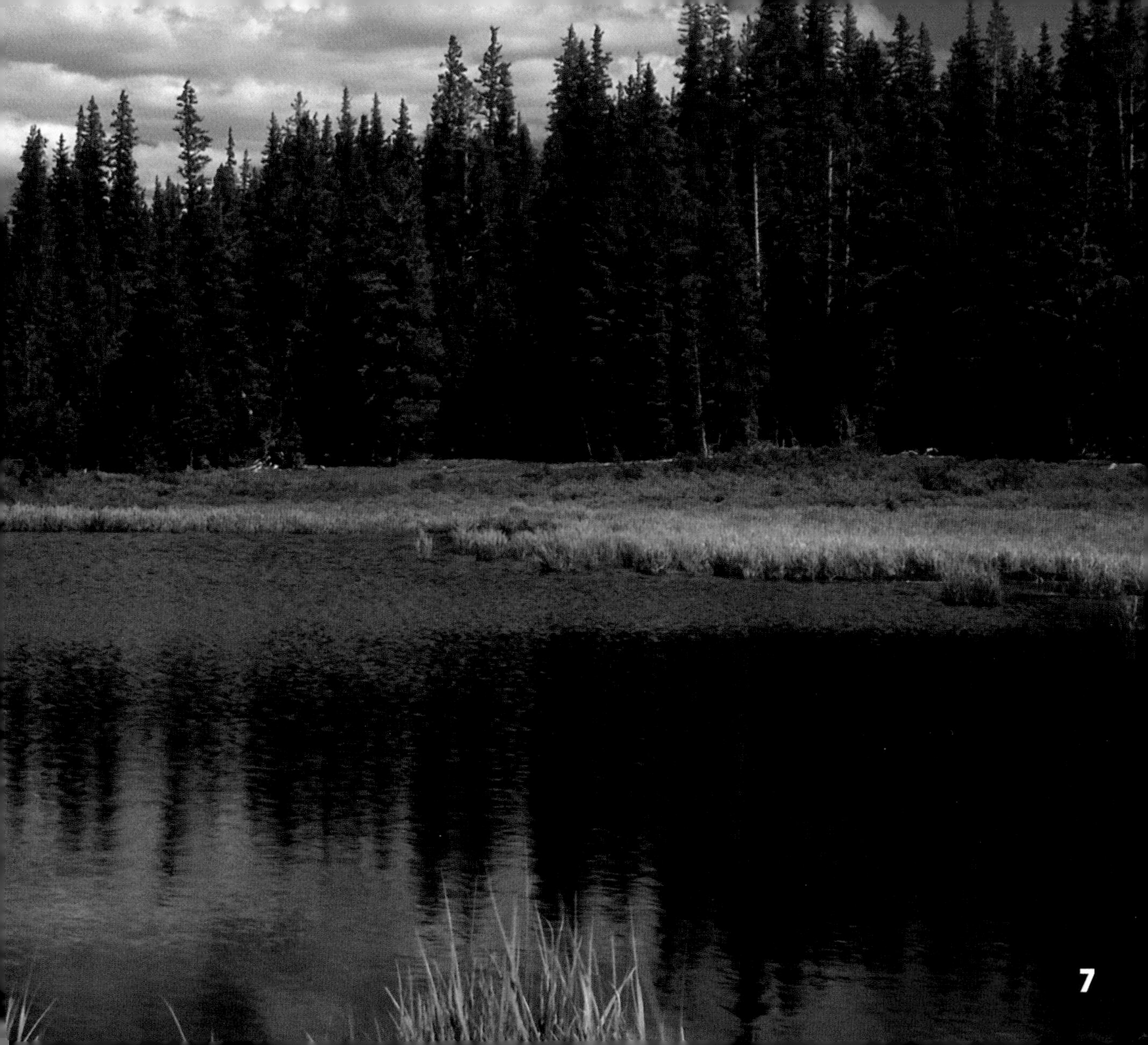

She likes to look at snails.

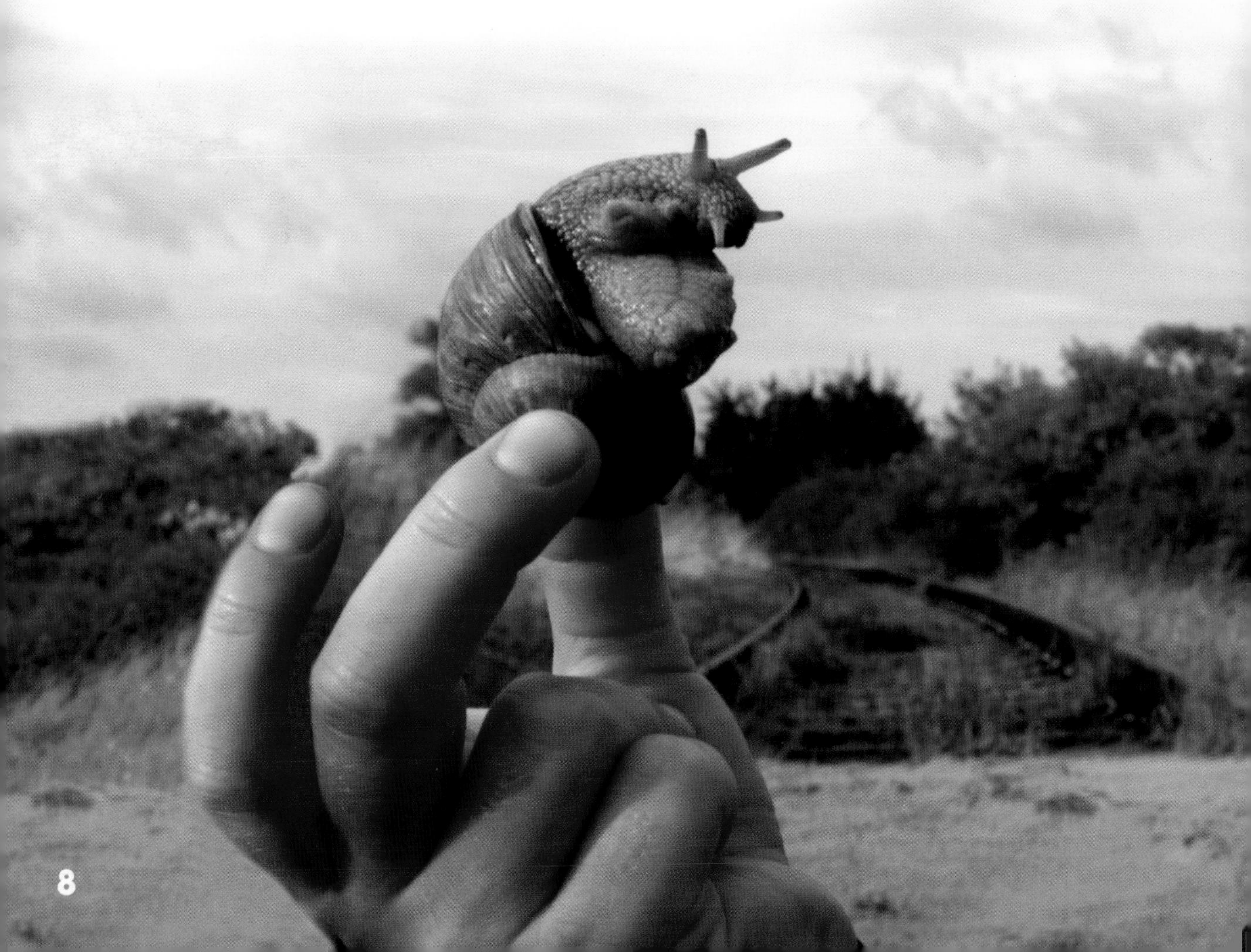

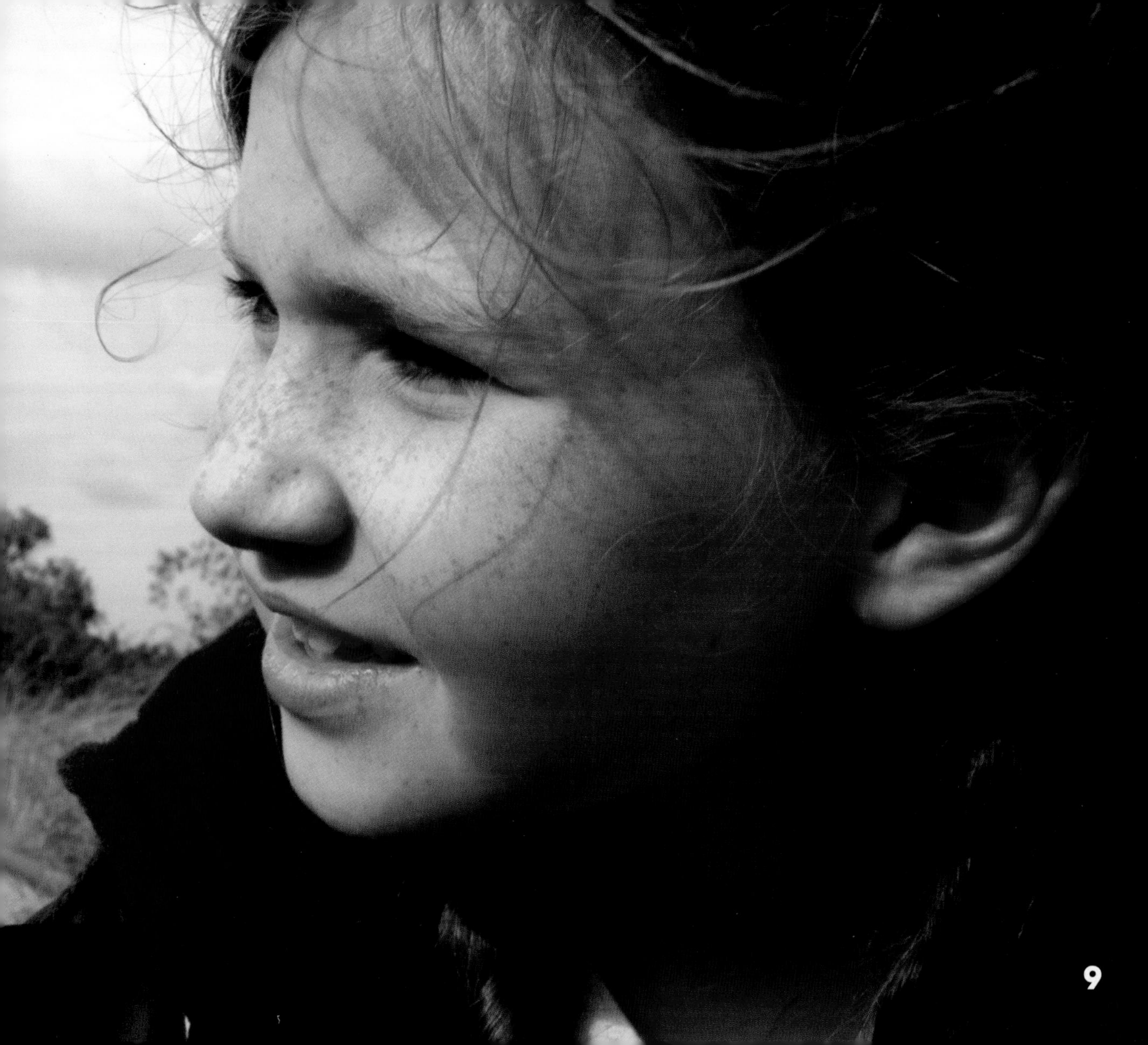

He goes to the park.

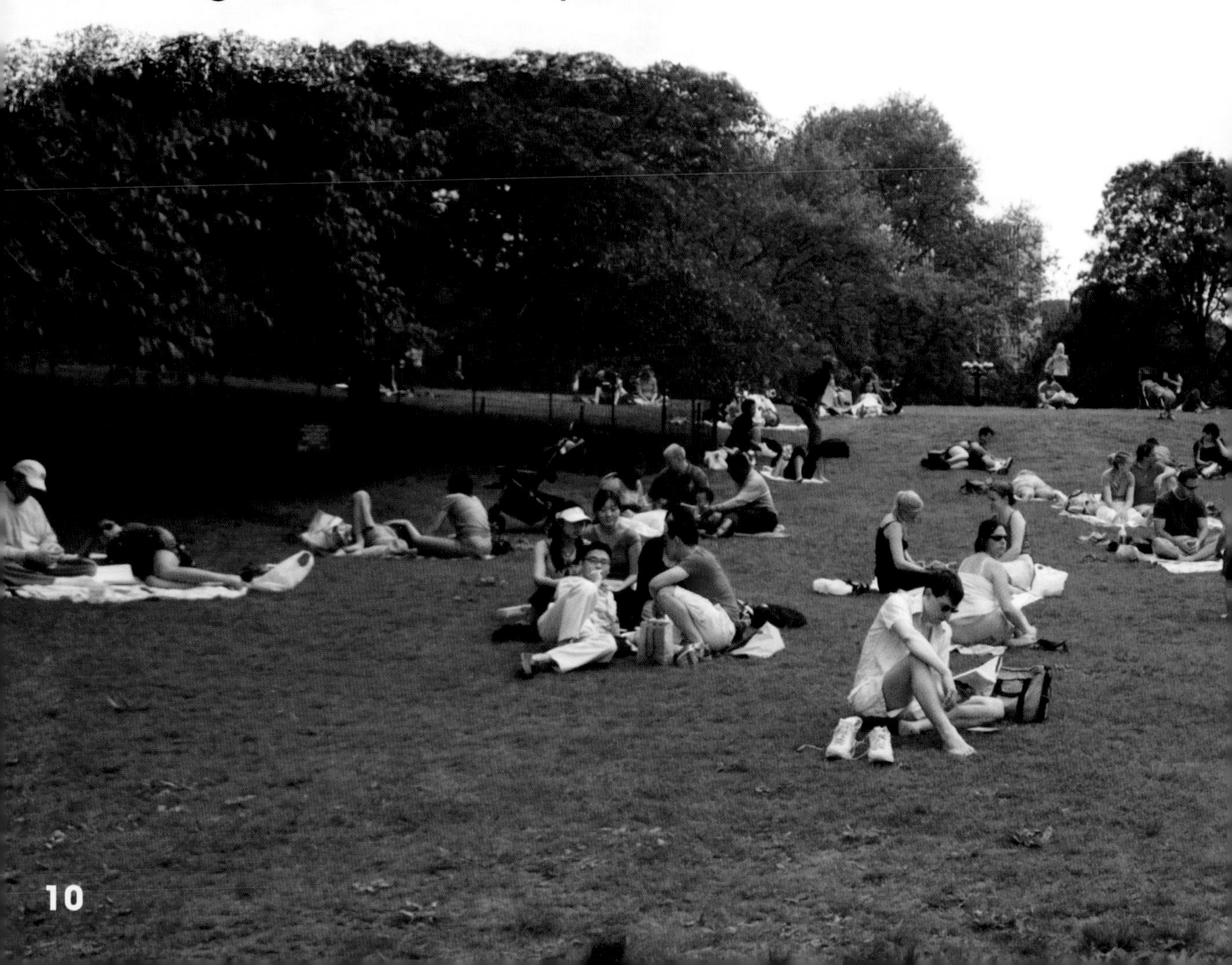

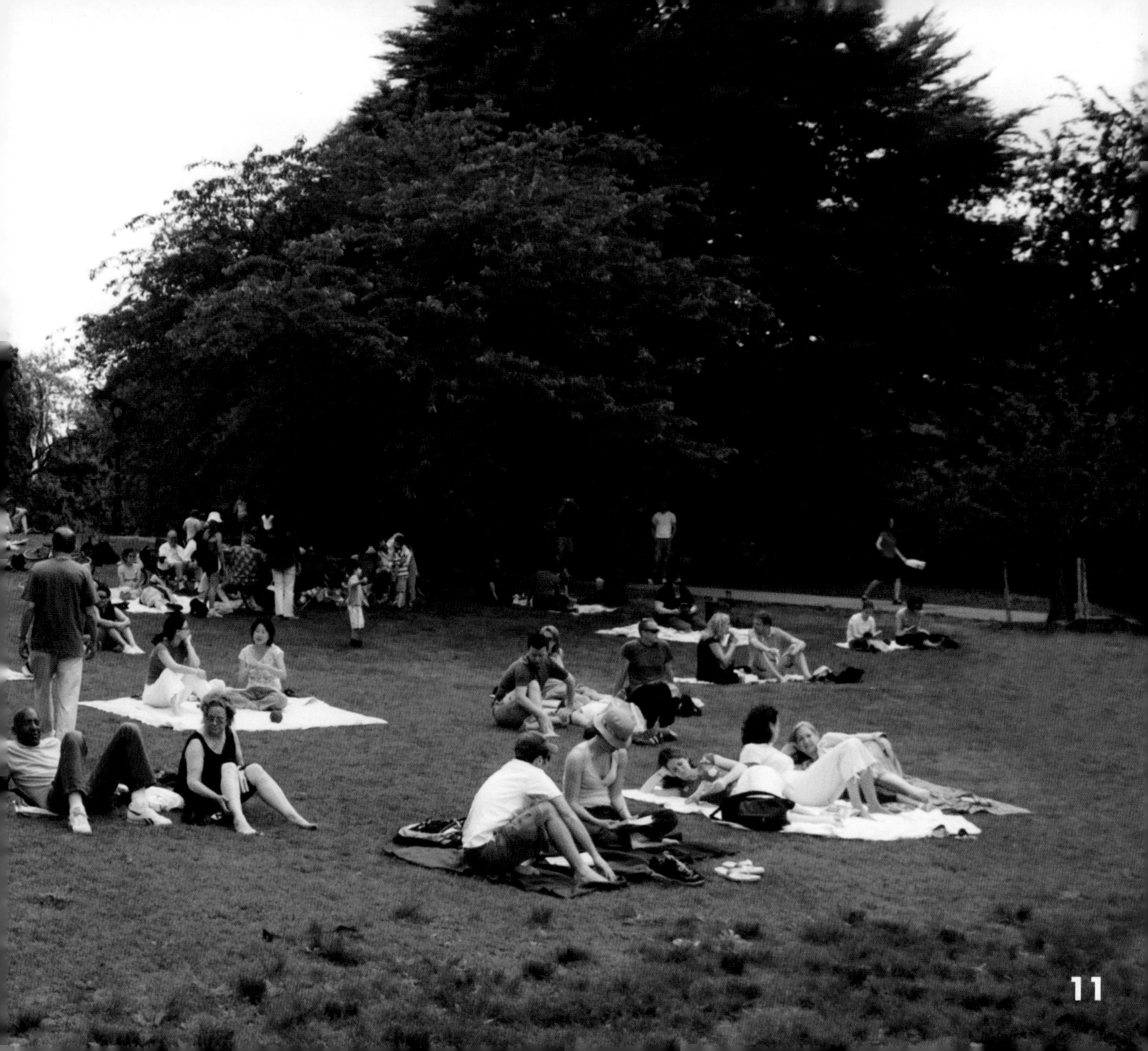

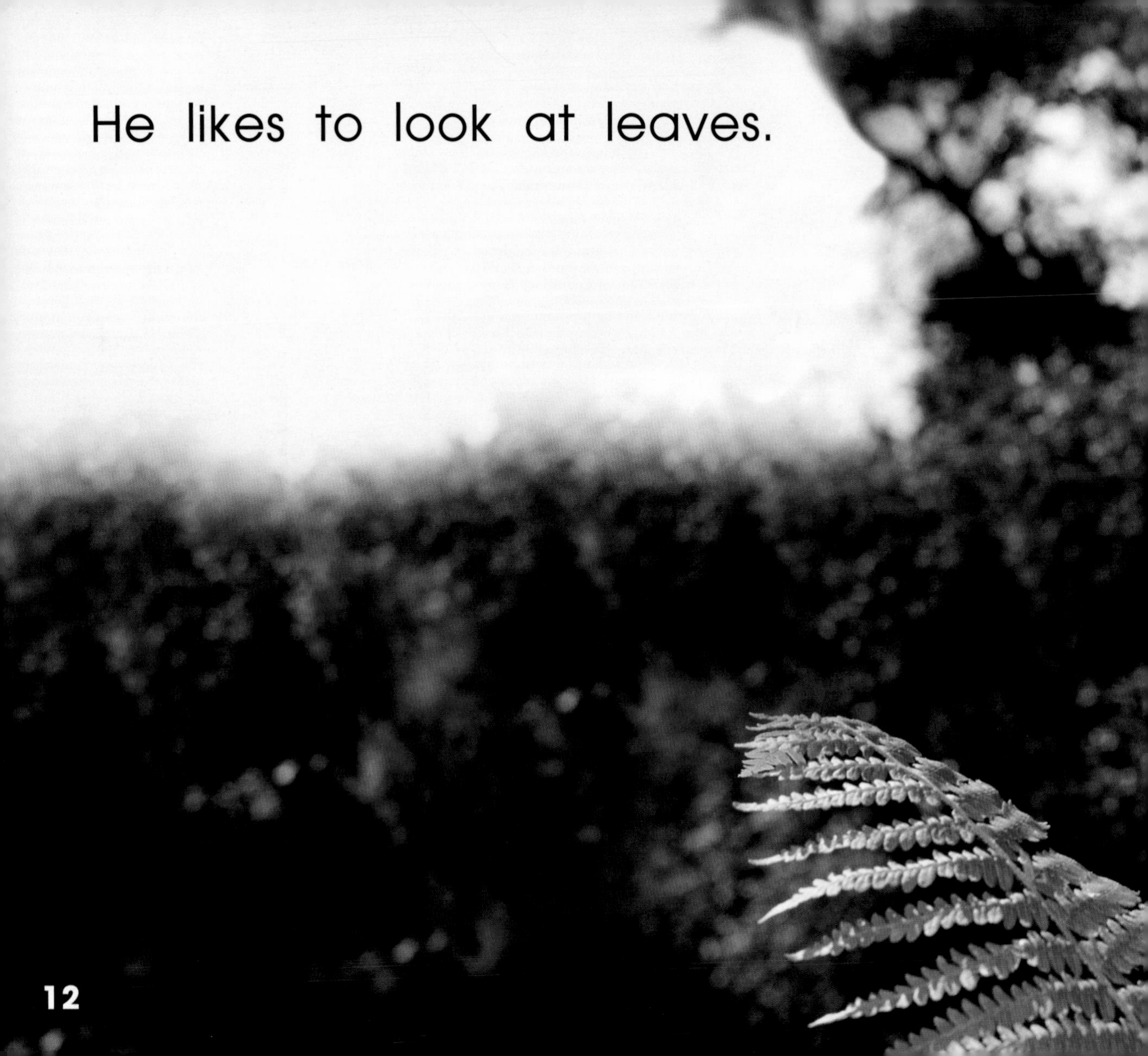

He likes to look at leaves.

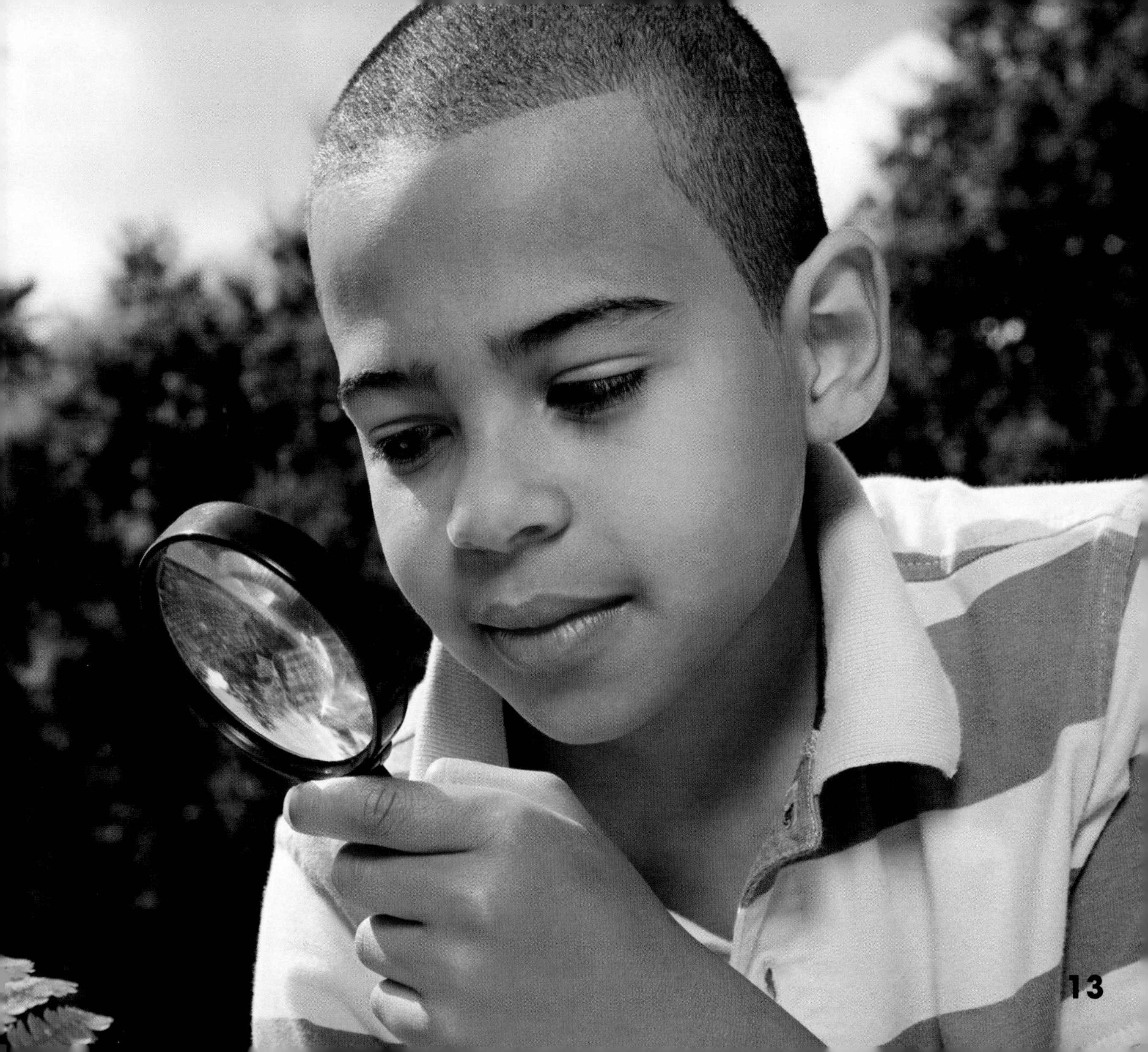

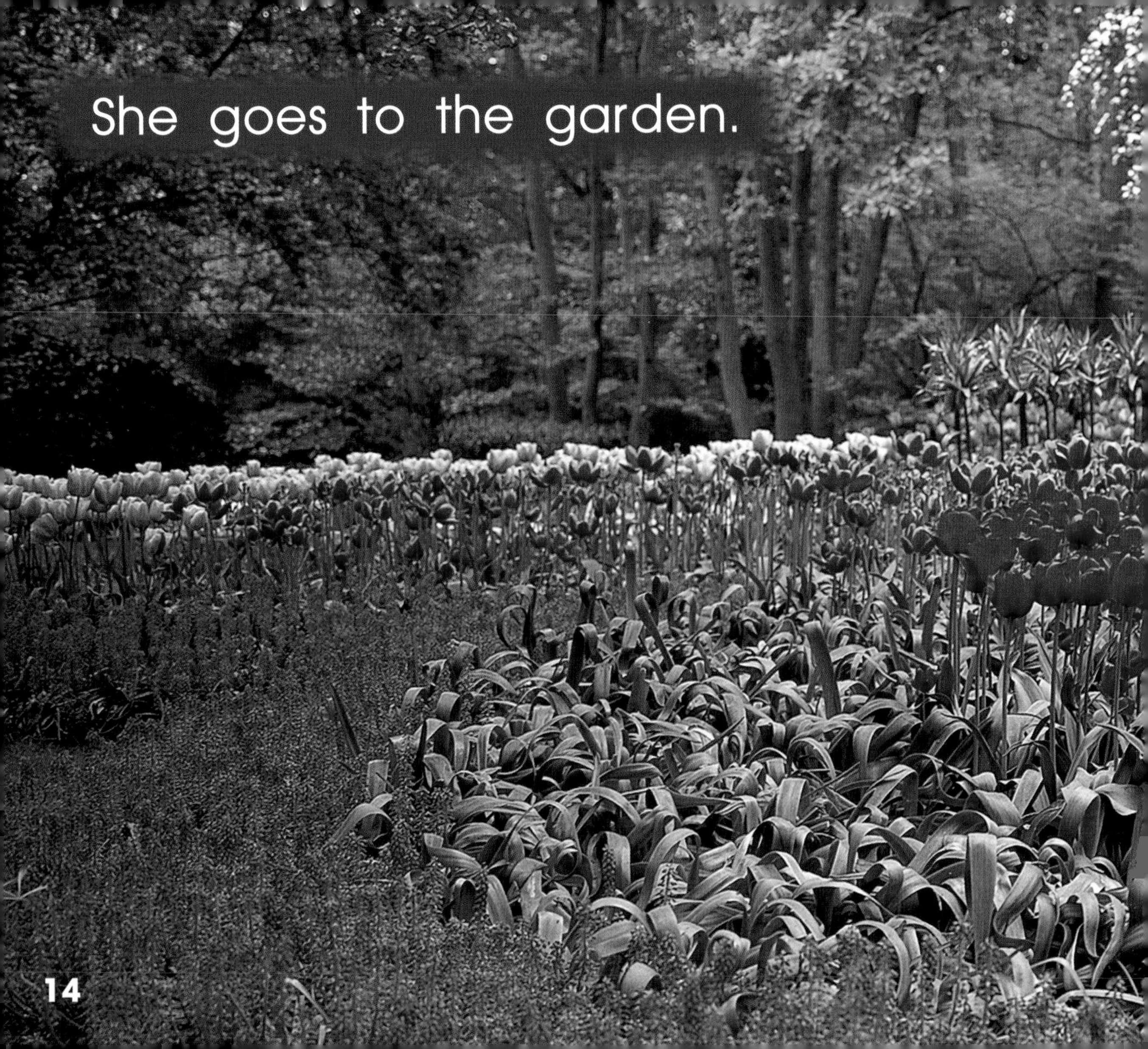

She goes to the garden.

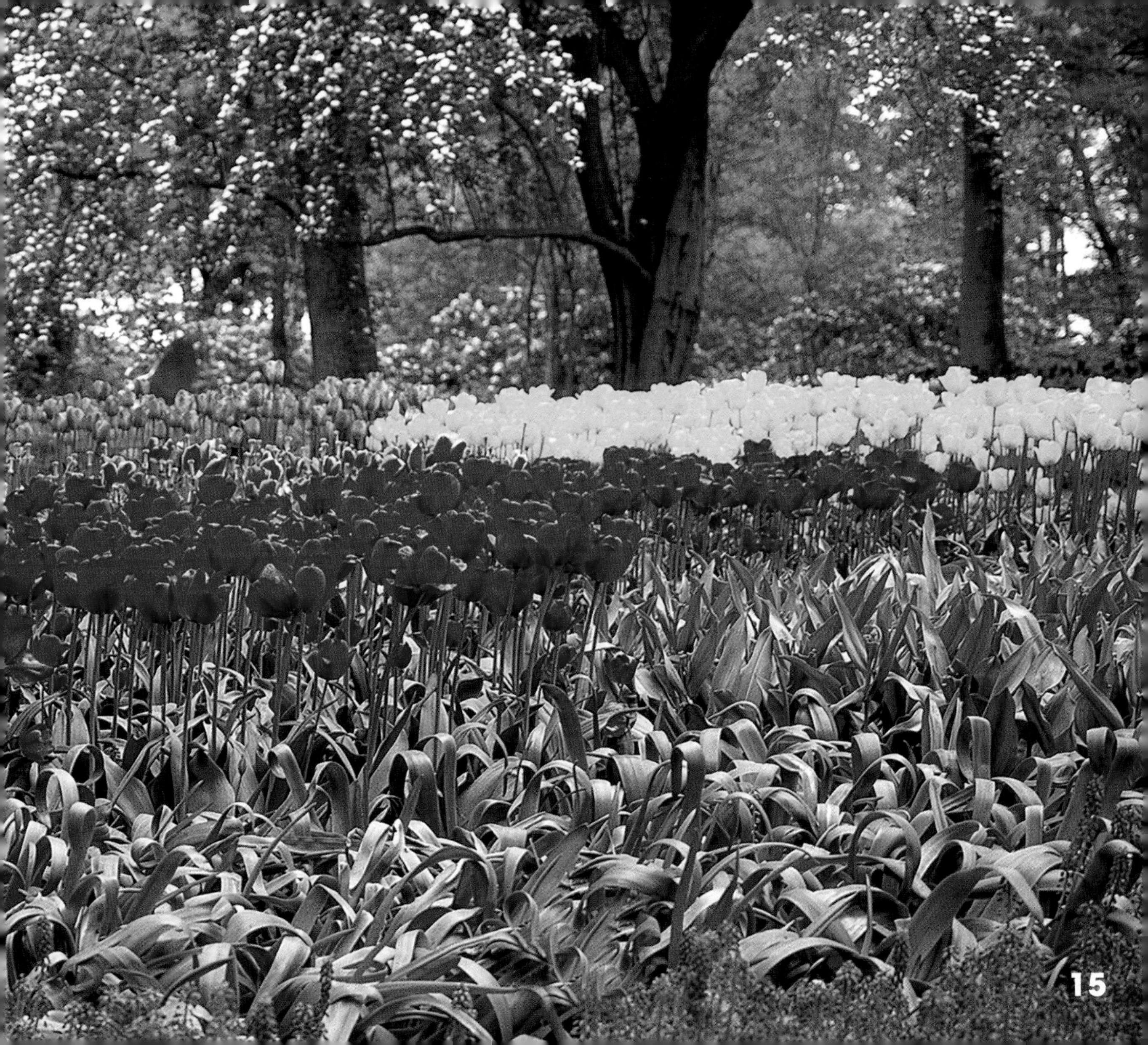

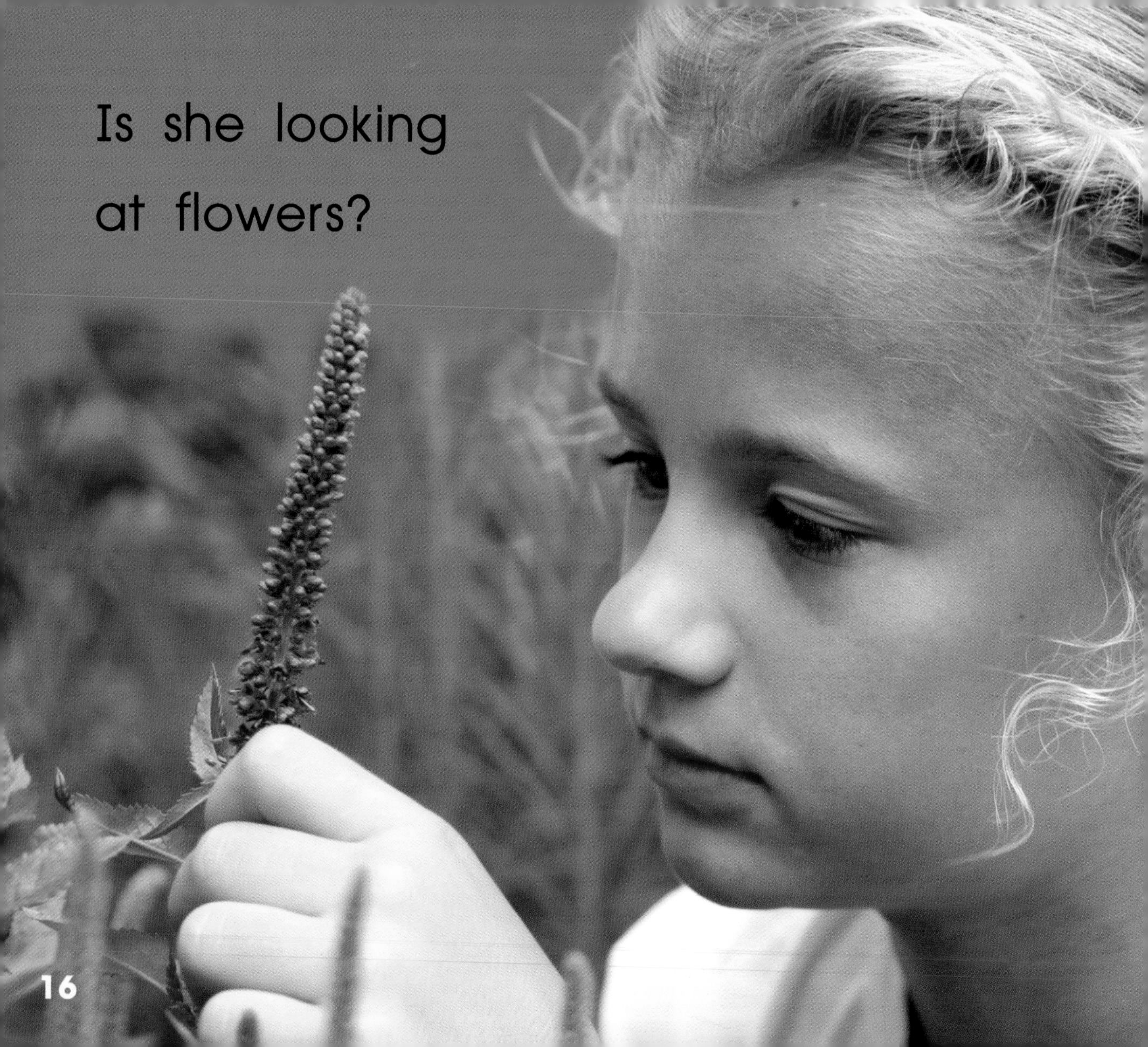

Is she looking at flowers?